Preface

Global warming is synonymous with climate change, rising sea levels, floods, and droughts, shifting rainfall patterns, acid rains, climate extremes, reduction in snow and ice, and clear and distinct possibilities of disbalanced ecosystems.

It is not something that has recently gained prominence, but has been right there in front of us. Evading the issue is not going to save us from the perils of global warming.

the Earth has been warmed so much that even a little extra warming may cause serious consequences to all forms of life – humans, plants, and animals. We can already witness that several species are turning extinct, landforms are changing, and the weather is turning unpredictable.

But this is not all, as there are several other dimensions to this issue. And we need to know about this subject in depth – causes, effects, science, and solutions. This book will in a simple easy way help you understand them.

Find out more about it through *50 FAQs on Global Warming*, so that you can also make a difference.

What's inside?

50 FAQs ON GLOBAL WARMING

teri

The Energy and Resources Institute

An imprint of The Energy and Resources Institute

First published in 2014 by
The Energy and Resources Institute
TERI Press
Darbari Seth Block, IHC Complex, Lodhi Road, New Delhi 110 003, India
Tel. 2468 2100/4150 4900, Fax: 2468 2144/2468 2145
India +91 ■ Delhi (0)11
Email: teripress@teri.res.in ■ Website: http://bookstore.teriin.org

ISBN 978-81-7993-452-4

Author: Himanshi Sharma
Publishing Head: Anupama Jauhry
Editorial and Production Teams: Rupak Ghosh, Ekta Sharma; Aman Sachdeva
Design and Illustration Teams: Santosh Gautam, Yatindra Kumar, Vijay Nipane
Image Researcher: Shilpa Mohan

Printed and bound in India

This book is printed on recycled paper.

Page No

What does the term "global temperature" mean?

Weather is the state of the atmosphere at a specific time in a specific place. Climate, on the other hand, is defined as the long-term weather patterns that describe a region. During the same month of the year, the temperature of India will differ from that of the UK, USA, or the UAE. This is because every country, state, city, or place has a different weather and climate. Some places are colder, while others are quite hot. We arrive at the global temperature by taking the average of all these temperatures.

1 Q

Is "global warming" same as "climate change"?

The two terms are not same, but related to each other. Global warming refers to the rise in average surface temperature worldwide due to emissions of greenhouse gases (GHGs) like carbon dioxide (CO_2). However, an increase in the proportion of CO_2 in the Earth's atmosphere will result in a lot of other changes besides higher temperatures. Collectively, these changes are known as climate change.

Thus, "global warming" implies rise in the Earth's surface temperature, while "climate change" is a broad term that includes global warming as well as everything else that will result from increasing amounts of GHGs in the atmosphere.

What is the greenhouse effect?

Greenhouses are made of glass and are designed to hold heat. Similarly GHGs, such as CO_2, water vapour, N_2O or nitrous oxide, and O_3 or ozone, trap the Sun's heat and do not let it escape. As a result, during the day, Earth's surface gets increasingly warmer. But when our planet cools down in the night, the heat is released back into the air. However, some heat still remains trapped in the lower atmosphere, keeping our planet warm and cozy. This natural process produced by GHGs is known as greenhouse effect – a term coined by Svante Arrhenius, a Swedish scientist.

3

Is the greenhouse effect good or bad for us?

Without the greenhouse effect, the Earth's average temperature would be about -18 ^{0}C, instead of the current pleasant 14 ^{0}C. In other words, our planet would be freezing, thus making it very difficult for most organisms, including humans, to survive. Therefore, the greenhouse effect is definitely good for us. But, if it gets too strong, then the Earth will keep getting hotter and hotter. Some human activities, such as burning coal and petrol, have added a lot of GHGs to the atmosphere. This has led to a marked increase in the Earth's temperature.

What are fossils fuels? How do they cause global warming?

Fossils are buried remains of living beings, including plants and animals, which existed millions of years ago. Fuels such as coal, petroleum, and natural gas are formed from these fossils and are known as fossil fuels. When fossil fuels are burnt to produce energy to run our cars and light up our homes, they release CO_2 and other GHGs. As a result, not only do these fuels pollute the environment but also raise global temperatures.

5

Do animals, too, raise the amount of greenhouse gases?
Yes, they do. Cattle, buffalo, sheep, goats, and camels produce methane – a GHG that is over 20 times more powerful than CO_2 over a 100-year period in trapping the Sun's heat. These animals release methane whenever they burp! During digestion, the bacteria inside their stomach decompose food, leading to the expulsion of methane. Moreover, when animal manure is stored or managed, methane gets released. At the global level, livestock accounts for about 18 per cent of total GHG emissions.

Is it true that the cultivation of rice can add to the greenhouse effect?

Rice plants require ankle-deep water to grow. This causes the organic matter in soil to ferment – a process that produces methane as a byproduct. The methane then escapes into the atmosphere through the roots and stems of rice plants. Being the world's second-most produced staple crop, rice cultivation is one of the largest manmade sources of methane emissions. As the world population keeps growing, global demand for rice will keep increasing in future, thus resulting in higher methane emissions. Such emissions from rice cultivation can be reduced through mid-season drainage and use of alternative fertilizers.

Do aircrafts contribute to greenhouse effect?

They do, by virtue of the fact that every time an aircraft takes off, it releases CO_2. In fact in 2012, aircrafts around the world were responsible for emitting a staggering 689 million tonnes of CO_2! The reason behind what is known as "air travel emissions" is that aircrafts run on fossil fuels like kerosene. Besides CO_2 (about 70 per cent), emissions from aircraft engines also contain about 29 per cent water vapour and less than 1 per cent of NO_X or oxides of nitrogen, CO or carbon monoxide, and soot, among others. It has been estimated that for every 220 km an aircraft flies, about 100 kg of CO_2 is emitted. To tackle the situation, an emissions limit has been suggested for aircrafts. Any aircraft that exceeds that limit will have to pay a carbon tax.

What is nitrous oxide?

It is a naturally-occurring GHG. However, because of various human activities, such as burning of fossil fuels, agricultural practices, wastewater management, and certain industrial processes (such as production of nitric acid), the proportion of N_2O in the atmosphere is on the rise. Not only do N_2O molecules remain in the atmosphere for 120 years on an average, but over a period of 100 years, N_2O, being a greenhouse gas, is 310 times more powerful than CO_2 as well!

Also known as laughing gas, N_2O was once used to treat patients with dental disorders. Dentists would use it to numb areas in the mouth. However, this practice is now banned in many places across the world. This is because studies have revealed that when it is released in the atmosphere, N_2O undergoes chemical reactions to form NO_x, which have ozone-destroying properties. Dentists now use local anesthesia in place of N_2O.

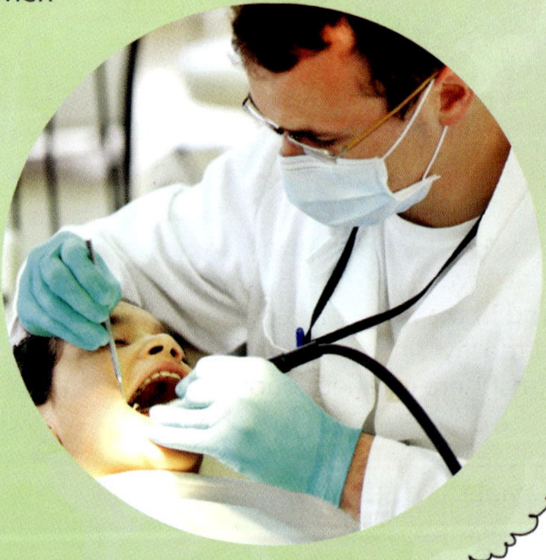

9

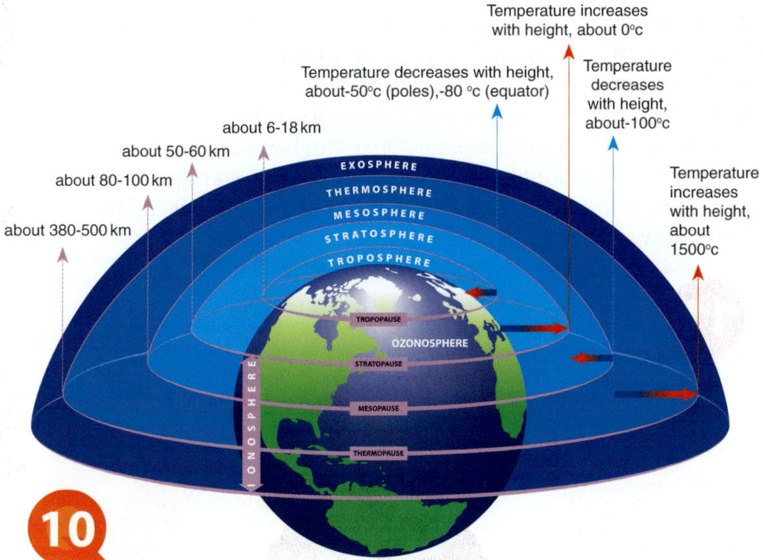

10

What is the ozone layer?

Ozone layer is formed of O_3 which is a naturally occurring gas found in the lowest layer of the Earth's atmosphere (troposphere) and the layer just above it (stratosphere). Ozone found in the troposphere is called "bad" ozone, because it causes smog. On the other hand stratospheric ozone is called "good" ozone, because it protects life on Earth by absorbing the harmful ultraviolet radiation from the Sun.

If the atmospheric ozone had not been present, with more and more ultraviolet rays from the Sun reaching the Earth, chances of us suffering from diseases, such as skin cancer, and eye damage, would have increased. Moreover, oceans, forests, and crops would have been badly affected.

Is the ozone layer thinning?
In the 1980s, scientists discovered that the ozone layer in a particular region above Antarctica had become extremely thin. Research revealed that certain chlorine-containing chemicals called chlorofluorocarbons (CFCs) were playing a major role in this thinning of ozone layer. Once the harmful impacts of CFCs were brought to the world's notice, it was decided to slowly phase CFCs out. Gradually, CFCs were replaced by hydrofluorocarbons (HFCs).

11

What are CFCs?

CFCs are chemicals, containing chlorine, fluorine, and carbon. Considered "miracle substances" because they are non-flammable, stable, low in toxicity, and cheap to produce, CFCs were used in manufacturing aerosol sprays and blowing agents for foam and packing materials, besides being used as solvents, and refrigerants.

During the late 1950s and early 1960s, CFCs made it possible for inexpensive air-conditioning in automobiles, homes, and offices. However, in 1974, CFCs were found to be the major source of chlorine in the stratosphere. Consequently, in 1987, 27 nations signed the Montreal Protocol to Reduce Substances that Deplete the Ozone Layer, which stipulated that the production of CFCs of these compounds should be reduced by 50 per cent of their 1986 production levels.

How do CFCs damage the ozone layer?

CFCs release chlorine when they come in contact with the Sun's ultraviolet rays. As a result of a number of chemical reactions, chlorine begins converting ozone into oxygen. We all know that oxygen is good for us, but not when it comes at the cost of ozone layer. In fact, one chlorine atom can destroy over 100,000 ozone molecules! This loss of stratospheric ozone results in the Sun's harmful ultraviolet radiation reaching the Earth's surface in higher amounts.

13

Chlorine is also released by natural sources. How can we be sure that this chlorine does not harm the ozone layer?

During a volcanic eruption, hydrogen chloride (HCL) gets emitted into the atmosphere. On the other hand, the source of chlorine in oceans is sea salt. Since most volcanic eruptions are not strong enough to reach the stratosphere, the emitted hydrogen chloride usually lands up in the troposphere. Similarly, sea salt is also released at a much lower level of the atmosphere. It has been estimated that in either case, the chlorine will have to be airborne for 2–5 years to reach the stratosphere. However, unlike CFCs, both sea salt and hydrogen chloride are easily soluble in water. So when rain falls, it effectively removes both these forms of chlorine from the troposphere.

What is the ozone hole?

The ozone hole is not really a "hole" but an area with reduced concentrations – less than 220 Dobson Units (unit for measuring ozone concentration) of ozone. It was discovered in 1984 by the British scientists Joe Farman, Brian Gardiner, and Jonathan Shanklin. They found that each spring, the ozone layer over Antarctica was thinning. This large, thin spot came to be known as the ozone hole. While this "hole" was as big as the USA, another smaller one appeared over the North Pole during the Arctic spring. Over the years, the ozone hole has steadily grown in size to reach 10.6 million sq. miles.

15

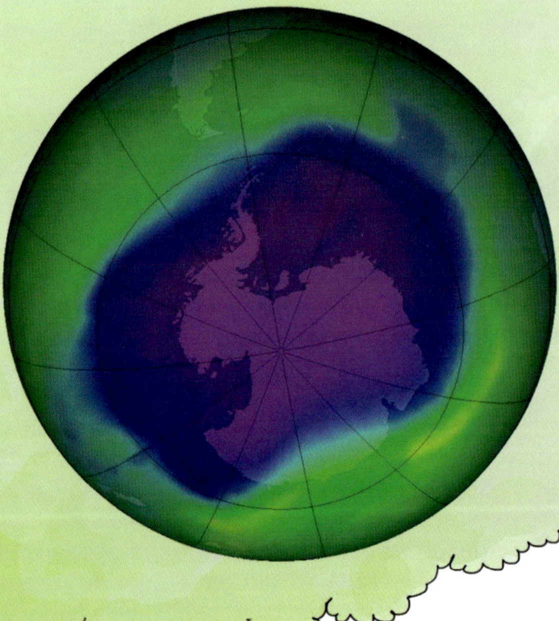

16

Can the lost ozone in the ozone layer be recovered?

O_3 is a naturally occurring gas, so we cannot produce it artificially. Moreover, certain CFCs, once released, can remain in the atmosphere for anywhere between 50 years and 100 years. So it will take several lifetimes for the lost ozone in the ozone layer to be recovered.

However, avoiding use of CFCs and other harmful gases, such as N_2O (emitted during fossil-fuel burning, application of agricultural fertilizers, and so on) can help. If such checks are applied on a large scale, scientists have predicted that the ozone layer might return to its original (pre-1980) levels sometime between the years 2060 and 2075.

What is the connection between 16 September and the ozone layer?

Worldwide, 16 September is observed as the International Day for the Preservation of the Ozone Layer. In 1995, the United Nations, an international organization dealing with issues of international concern, chose 16 September to commemorate the date of signing of the Montreal Protocol on Substances that Deplete the Ozone Layer. On this day, every year, all member states of the UN are encouraged to promote concrete activities that deal with the protection of the ozone layer.

18

What are aerosols? How do they impact the temperature on Earth?

Tiny solid particles or liquid droplets floating in the air are called aerosols. Their size varies from a few nanometres to a few micrometres (1 nanometre = 1 billionth of a metre; 1 micrometer = 1 millionth of a metre).

There are two types of aerosols – natural and manmade. Forest fires, clouds, dust storms, volcanoes, desert dust, and sea spray emit natural aerosols, while cigarette smoke, fumes from cooking stoves, burning candles, and hairspray cans emit manmade aerosols.

Aerosols are capable of trapping the Sun's radiation and causing the greenhouse effect. However, aerosols are not all that bad because they may lead to a cooling effect as well. This happens in two ways: through reflection of solar radiation under clear sky conditions and by increasing the reflectivity of clouds and making them last longer.

Besides adding CO_2 to the atmosphere, is there any other way burning coal harms the environment?

A byproduct of burning coal in thermal power plants is fly ash, a fine powdery material. Studies reveal that fly ash carries into the atmosphere 100 times more radiation than a nuclear power plant producing the same amount of energy! Fly ash also contains trace amounts of arsenic, lead, and cadmium – all of which are harmful to human health. India produces about 112 million metric tonnes of fly ash every year, which is dumped in storage ponds and landfills. Consequently, it hampers soil fertility and contaminates both soil and groundwater.

19

What caused the "Great London Smog"?

Also known as The Big Smoke, this event occurred in London during December 5-9 in 1952. To keep their homes warm during an uncommonly freezing winter, citizens of London began to burn more coal than usual. This caused a lot of smoke, which mixed with the cold air over the city to form the densest smog London has ever seen. The thick layer of smog blocked sunlight and caused breathing problems that killed about 12,000 people in a matter of months.

What is "acid rain"?

The term "acid rain" was coined in 1872 by Robert Angus Smith, an English scientist. When pollutants like N_2O and sulphur dioxide (SO_2) are emitted, sometimes they combine with water vapour in the atmosphere to form nitric acid and sulphuric acid. Acid rain refers to a mixture of both wet and dry materials deposited from the atmosphere that contains higher than normal amounts of nitric and sulphuric acids. Wet deposition includes acid rain, fog, and snow, while dry deposition can get incorporated with dust or smoke and fall on the ground. Acid rain is harmful for aquatic plants and animals. It corrodes buildings and car paint and is also bad for humans, because it can lead to respiratory problems and allergies.

21

22

Does acid rain escalate the rate of global warming?

Acid rain weakens trees by damaging the waxy coating of leaves, washing away the nutrients available to plants, and by exposing them to poisonous substances released from the soil. It causes severe damages to forests, which, in turn, means less production of O_2 or oxygen and more production of CO_2. Excess of CO_2 in the air directly translates into higher temperatures and consequently an accelerated rate of global warming.

Is there any forest that has been particularly affected by acid rain?

Famous for its pine and fir trees, the Black Forest is located in southwest Germany. It is believed that it was in this forest that Hansel and Gretel met the wicked witch. In the 1960s, scientists first noticed that the trees had begun losing up to a quarter of their leaves and needles in comparison to preceding years (decline in leaves is one of the most obvious effects of acid rain on plants). By 1990, about half the trees in the forest had exhibited this disturbing trend.

On the other hand, seedlings were adversely affected by the highly acidic soil. Gradually, the needle-less trees began to look dead and black. Germans called it *Waldsterben* (tree death).

23

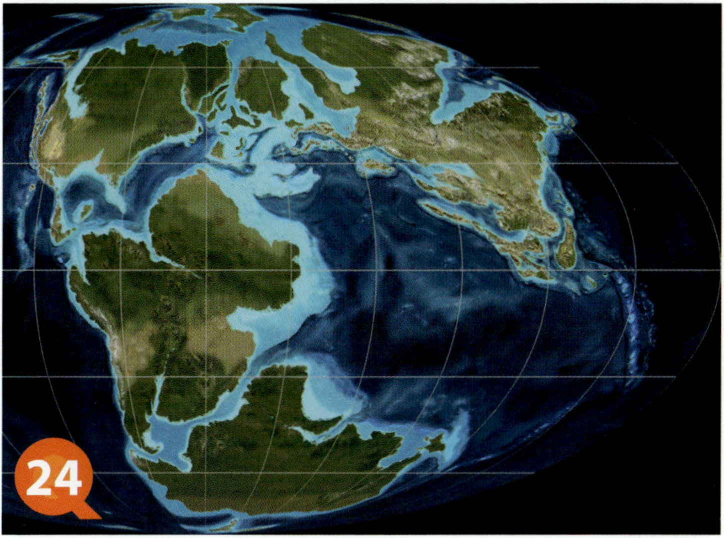

24

What is continental drift and how does it affect the climate?

In the beginning, the entire landmass on Earth was one huge chunk, known as Pangaea. With the passage of time, Pangaea broke into separate continents that slowly started moving away from one another. This phenomenon is called continental drift. The drift also had an impact on the climate because it led to changes in the physical features and position of the landmasses, as well as the positions of the waterbodies. The separation of the landmasses changed the flow of ocean currents and winds, thus impacting the climate.

Continental drift is occurring even today. For instance, the Himalayas are rising by about 1 mm annually, because the landmass that is India is slowly but steadily moving towards the landmass called Asia.

Which areas of the world will be affected the most by global warming?

The "global" in "global warming" implies that the entire world will be impacted. People living in low-lying coastal areas will face frequent flooding that will damage crops, spread diseases, and destroy homes. It will also cause deserts to expand, thus causing food scarcity. As the seas get warmer, devastating hurricanes will occur more frequently in the Atlantic. Higher temperatures will also cause a rise in the intensity and ferocity of forest fires in places with dry climate.

The 2012 edition of *Climate Change and Environment Risk Atlas* ranks 193 countries on the basis of their vulnerability to climate change, of which 30 countries were at extreme risk. The top 10 were: Haiti, Bangladesh, Sierra Leone, Zimbabwe, Madagascar, Cambodia, Mozambique, Democratic Republic of Congo, Malawi, and the Philippines.

25

What does carbon sequestration mean?

Carbon sequestration refers to the process of removing carbon from the atmosphere. The natural process occurs as storage of carbon in trees, plants, fossil fuels, or some other medium.

Plants store carbon which can be found in fossils and wood. If burnt, this carbon will add to the existing atmospheric carbon, thereby leading to further global warming, and subsequently climate change. Trees and plants not only keep our surroundings clean and green, but also save a great deal of carbon from escaping in the atmosphere and harming the environment.

There are other forms of carbon sequestration in nature, occurring in soil and oceans. Sequestration through human-made efforts, using high end technologies, is being explored in the form of geo-engineering or as carbon capture and storage (CCS).

Has any island already disappeared by rising water levels?
A tiny island in the Bay of Bengal, which was called New Moore Island in India and South Talpatti in Bangladesh, has been completely submerged under the rising sea. In 1996, another nearby island called Lohachara also disappeared, forcing its inhabitants to shift to the mainland. At least 10 other islands of the region might soon meet the same fate.

27

28

In what ways are polar bears being affected by global warming?

Polar bears are marine mammals that spend a lot more time on sea than on land. Due to global warming, the Arctic ice is melting. This is where the polar bears hunt for food. So hunting opportunities for polar bears have decreased. As a result, they are faced with food scarcity. All this has led to a drastic reduction in their population. In fact, scientists are of the opinion that if the current warming trend continues, then the polar bears will be on the verge of extinction by the next century. Besides habitat loss, the retreat of Arctic ice has also widened the gaps of open water between the ice and land, thus compelling these animals to swim from shore to ice, sometimes leading to drowning. The shrinking Polar ice cap has also led to a fall in the number of seals, which is food for polar bear. As a result, these bears are now forced to stay hungry for long periods of time.

Is any particular fish in danger due to global warming?

Trout and salmon are two cold water fish that are at severe risk due to rising temperatures. They cannot live in warm water, since the ideal temperature for them to survive is 10–18 °C. Global warming has resulted in rising water temperatures all over the world. A study by the US-based Natural Resources Defense Council (NRDC) has revealed that by the year 2090, global warming is likely to cause the disappearance of trout and salmon from 18–38 per cent of their current habitat.

29

30

How are hurricanes related to global warming?
Scientists suggest that hurricanes and other tropical storms are now taking nine hours lesser to reach Category 3 speed (wind speed of 178–208 kmph) than the time they took just 25 years ago. The reason is believed to be rising temperatures of oceans, which have risen by 0.5 °C across the globe. Warmer the ocean's surface, greater the energy to power a storm. A group of international researchers has found that for every 1 °C rise in temperature, the number of hurricanes in the Atlantic that are at least as strong as – or stronger than – Hurricane Katrina will rise anywhere between two-fold and seven-fold!

However, it should be pointed out that besides global warming, there could be other environment-related reasons such as natural climate cycles that can have the same effect on hurricanes.

Can forests and oceans help in controlling global warming?

Yes, they can! Trees and plants convert the CO_2 present in air to oxygen. In other words, they act as carbon "sinks" (or storehouses). As forests contain a huge number of plants and trees, they can play a major role in cleaning the air in a relatively short period of time. A study conducted by the University of Leeds has revealed that every year existing forests absorb nearly 8.8 billion tonnes of CO_2. An additional 6 billion tonnes are absorbed by newly planted forests around the globe.

Similarly, oceans too absorb CO_2. According to a study conducted jointly by the World Ocean Circulation Experiment (WOCE) and the Joint Global Ocean Flux Study (JGOFS), the oceans of the world have absorbed 48 per cent of all CO_2 emitted as a result of fossil-fuel combustion and cement manufacture between 1800 and 1994.

31

What was the Chipko Movement?

Also known as the *Chipko Andolan*, it was an environmental movement that started in the 1970s. *Chipko*, meaning "to cling or stick to something", is a Hindi word. On 26 March 1974, a group of rural women from the Chamoli district in Uttarakhand hugged trees and insisted that they would prefer to be axed first before allowing the trees to be cut down. It was their way of protesting against the felling of trees.

Supporters of the Chipko movement, majority of which are village women, have been successful in banning the cutting of trees in several regions of the country – such as the Western Ghats and the Vindhyas – and have also played a major role in formulation of the natural resource policy of the country.

How are coral reefs affected by increased amounts of CO_2?

Coral reefs absorb CO_2, thus acting as a natural carbon sink. Moreover, globally they offer shelter to over 4,000 species of fish, crabs, starfish, and turtles. However, if too much of CO_2 finds its way into the water, the oceans become acidic. A research led by scientists from Stanford University suggests 9,000 coral reefs around the world will collapse by the year 2100 when the atmospheric CO_2 levels are expected to reach 560 parts per million (as of May 2013, the concentration stood at 399.89 ppm). Alongside the corals, the hugely diverse marine life in the oceans will also be drastically reduced.

33

How are corals affected by global warming?

Tiny algae, called zooxanthallae, live inside coral tissues. While the coral provides the algae with a home, the algae, in turn, offer the coral up to 95 per cent of its food. However, due to global warming, when the ocean's temperature becomes too hot, corals expel the algae. This phenomenon is called coral bleaching. It leads to a steady loss of colour, with many coral colonies turning completely white. Coral bleaching has been reported throughout Southeast Asia, Indian Ocean, and Pacific Ocean.

Which country emits the maximum amount of CO_2?

According to the *Trends in Global CO_2 Emissions 2012 Report*, China is the world's largest emitter of CO_2. In 2011, it emitted an estimated 9.700 million tonnes of CO_2. In the second place was the USA (5,420 million tonnes), followed by India (1,970 million tonnes). However, when it came to maximum CO_2 emissions per capita, Australia (19 tonnes per head) fared the worst, followed by the United States (17.3 tonnes), and Saudi Arabia (16.5 tonnes of CO_2 per capita). It might be mentioned here that China's per capita emissions of CO_2 were only 7.2 tonnes, while India recorded an even smaller figure (1.6 tonnes of CO_2 emissions per capita).

35

36

What is CNG? Why is it considered environment-friendly?

Natural gas is a combustible mixture of hydrocarbon gases. Its major constituent is methane, but it can also include ethane, propane, butane, and pentane. Since the 1930s, it has been used as a fuel for vehicles. As an alternative fuel, it can be used either in the form of compressed natural gas (CNG) or liquefied natural gas (LNG).

CNG is clear, odourless, and non-corrosive. It has been used as a vehicle fuel in place of petrol and diesel since the 1960s. There are three reasons that make CNG environment-friendly: it is the cleanest burning fuel, generates lesser emissions, and produces significantly less pollutants than petrol or diesel.

What is carbon footprint?

Carbon footprint is a measure of the impact a person or group has on the environment. It is measured in terms of the GHGs generated while carrying out various daily activities. The impact is measured in units of CO_2 – the commonest of all GHGs.

Carbon footprint is of two types: primary and secondary. Primary footprint includes emissions from the things we do (for example, driving a car). On the other hand, secondary footprint measures the CO_2 emitted by others while manufacturing the goods or providing the services that we buy (for instance, the clothes we wear and the food we eat). Carbon footprint is usually expressed in tons of CO_2-equivalent per year.

37

What is meant by CO_2-equivalent?

Carbon dioxide equivalent is a unit of measurement used to compare emissions from various GHGs based on their global-warming potential (global-warming potential of a GHG is defined as the ratio between heat trapped by a unit mass of the gas and that of a unit of CO_2 over a specific time period).

CO_2-equivalent is generally expressed as million metric tons of carbon dioxide equivalents. It is derived by multiplying the tons of a gas by the associated global warming potential. For example, the global warming potential for methane over 100 years is 21. This means that over a 100-year-period, emissions of 1 million metric tons of methane are equivalent to emissions of 21 million metric tons of CO_2.

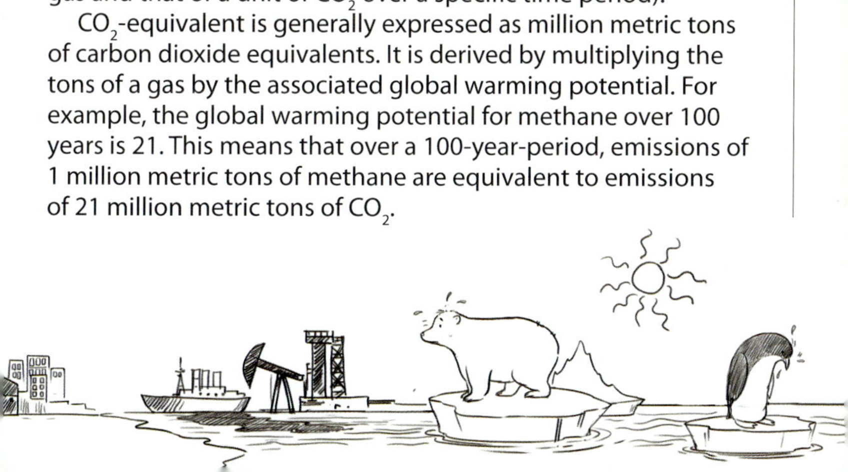

What is the Montreal Protocol?

The Montreal Protocol on Substances that Deplete the Ozone Layer is an international treaty aimed at protecting the ozone layer. It stipulated that the member states would phase out the production of substances that are responsible for ozone depletion. The Protocol was opened for signature in Montreal, Canada, in 1987 and came into force in 1989. If it is observed seriously, the depleted ozone layer is expected to return to its pre-1980 levels sometime between the years 2060 and 2075.

39

What is the Kyoto Protocol?

It is an international agreement for limiting GHG emissions. Its importance lies in the fact that it is the first-ever legally binding contract to reduce GHGs at the global level. The Protocol was adopted in Kyoto, Japan on 11 December 1997 and came into force on 16 February 2005. The agreement specifies targets for each member country regarding reduction in the emissions of six GHGs, including carbon dioxide (CO_2), methane (CH_4), nitrous oxide (N_2O), hydrofluorocarbons (HFCs), perfluorocarbons(PFCs), and sulphur hexafluoride(SF_6).

Which important mechanism was introduced by the Kyoto Protocol?

It is CDM, or Clean Development Mechanism, which was defined in Article 12 of the Protocol. CDM is an interesting way to check GHG emissions. As part of this mechanism, industrialized countries like the USA can invest in emissions-reduction projects in developing countries (like India). This will serve two benefits. Such projects will allow industrialized countries some flexibility in meeting their emissions-reduction targets. At the same time, they will help the developing nations in reducing their emissions.

41

42

How is the Kyoto Protocol related to the year 2012?

The member countries of the Kyoto Protocol all agreed to what is known as the "first commitment period" to reduce emissions of GHGs. This period expired at the end of 2012. In the Climate Summit in Doha towards the end of 2012, the Protocol was extended for a second commitment period to 2020. Countries will now be allowed to review their emissions targets, but they will not be able to cut down upon the targets set, and will only be able to strengthen them.

What is meant by "emissions trading"?

Emissions trading system (ETS) involves trading of rights to emit pollutants such as CO_2 and SO_2. Developed in the 1970s and 1980s, ETS was introduced in the United States in 1990 to combat acid rain. However, it has quickly evolved to become a way of tackling GHG emissions linked to climate change.

Under the ETS, a strict emissions limit (called credits) is set for all countries stipulating the amount of CO_2 they can generate. Countries that exceed the limit have the option to buy credits from countries with lesser emissions. In this way, developed countries are encouraged to lower their emissions, while developing countries that emit lesser emissions can earn money by trading their credits.

43

What is the UNFCCC? What is its objective?

The United Nations Framework Convention on Climate Change (UNFCCC) is an international treaty on environment. It was opened for signature on 9 May 1992 and was enforced in March 1994.

The objective of the treaty is to stabilize the concentrations of GHGs in the atmosphere at a level that would lessen their harmful impacts on the climate. When it was framed, the UNFCCC did not specify any limits on GHG emissions for countries, nor did it have any provisions of enforcement. In their place, it included provisions for updates, called protocols, which would set mandatory limits for emissions. A major update was the Kyoto Protocol.

What is meant by COP?

Countries signing the UNFCCC are known as Parties. The meeting of the Parties to assess progress in matters related to climate change is known as the Conference of the Parties, or COP. From 1994 to 1996, the COP was organized annually. However, following a change of rules in 2010, it was decided that COP will be held every two years.

45

46

What is the polluter pays principle?

An important element of the UNFCCC is the polluter pays principle (PPP). It stipulates that those who cause pollution should bear its costs. For example, if a factory emits a harmful substance as a byproduct of its operations, then it should be responsible for the safe disposal of the substance. Specifically, the PPP has been applied to emissions of GHGs.

Usually, PPP is implemented through two policy approaches – command-and-control and market-based. While the former includes performance and technology standards, the latter comprises pollution taxes, tradeable pollution permits, and so on.

What is The Earth Summit?

The Earth Summit is the popular name for the United Nations Conference on Environment and Development (UNCED) held at Rio de Janeiro, Brazil between 3 and 14 June 1992. As many as 172 governments and 108 heads of state or government participated in The Earth Summit, which had environment and sustainable development as its principal theme. The major documents that evolved out of the Summit were Agenda 21, the Rio Declaration on Environment and Development, and Forest Principles.

Agenda 21 is a comprehensive plan of action to be undertaken at the global, national, and local levels by organizations belonging to the United Nations system, governments, and important groups in every area in which human activities affect the environment.

47

What is the IPCC? What kind of work does it do?

The IPCC, or the Intergovernmental Panel on Climate Change, is an international scientific organization. It was formed in 1988 to study the scientific, technical, and socio-economic information for understanding climate change. It also assesses the harmful impacts of climate change and looks for viable solutions.

The scientific evidence presented by the first IPCC Assessment Report of 1990 played a major role in the formation of the UNFCCC. Since then, the IPCC has regularly come up with extremely comprehensive Assessment Reports on climate change. Currently, 195 countries are IPCC members. Under the chairmanship of Dr R K Pachauri, the IPCC won the Nobel Peace Prize in 2007 for its work on climate change.

How is the Indian government tackling the problems of global warming and climate change?

To meet the challenge of climate change, on 30 June 2008, the National Action Plan on Climate Change (NAPCC) was released by the Indian government. This policy document prepared by the Prime Minister's Council on Climate Change necessitated the establishment of eight national missions.

These missions are: National Solar Mission, National Mission for Enhanced Energy Efficiency, National Mission on Sustainable Habitat, National Mission on Water, National Mission for Sustaining the Himalayan Ecosystem, National Mission for a Green India, National Mission for Sustainable Agriculture, and The National Mission for Strategic Knowledge for Climate Change. Each mission falls under a specific ministry.

49

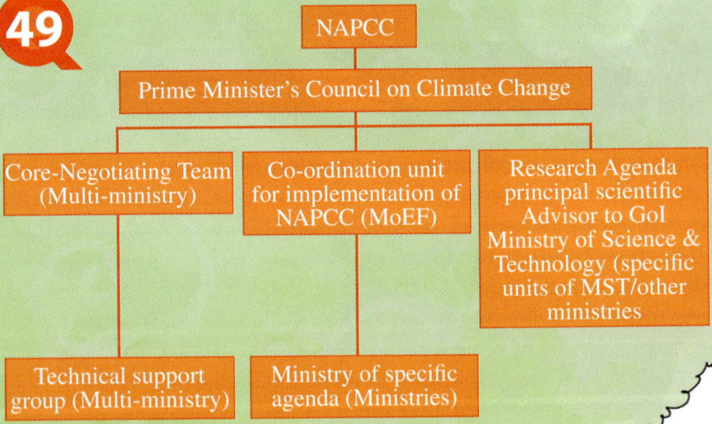

What can I do to control global warming?

- Buy energy-efficient appliances, turn off lights while leaving a room, unplug all appliances when they are not in use, run washing machines with full load, operate air-conditioners in the economy or energy-saving mode.
- Reduce petrol usage, use public transport.
- Pay electricity, mobile, and credit card bills online.
- Say "No" to plastic bags. For your next trip to the local market, carry a cloth bag, instead.
- Recycle and reuse paper. Print a document only when absolutely necessary, and then, too, print on both sides.
- Turn the tap off while brushing your teeth, fix all leaking faucets, install water-efficient appliances.
- Practise the 3Rs: Reduce the amount of new stuff you buy, reuse the stuff you already have, recycle whatever you can, from bottles and cans to jars, magazines, and newspaper.

Glossary

Acid rain precipitation that contains higher-than-normal amounts of nitric and sulphuric acids

Aerosol suspension of fine solid or liquid particles in gas

Algae tiny, plantlike water-based life forms

Carbon footprint amount of greenhouse gases (usually carbon dioxide) emitted directly or indirectly as a result of human activities within a given time frame (usually a year)

Carbon sink a natural reservoir (such as forests or oceans) that absorbs the CO_2 released into the atmosphere during burning of fossil fuels

Climate change a change in global climate patterns caused primarily by the higher levels of CO_2 in the atmosphere

Climate average condition of the weather at a place over a period of years

Continental drift gradual movement of the continents across the Earth's surface over time

Coral a hard material formed on the bottom of the sea or ocean by the skeletal remains of tiny creatures called polyps

Ecosystem everything that exists in a particular environment, including living things (such as plants and animals), as well as non-living things (such as rocks and soil)

Emissions trading a market-based approach to reduce emissions of greenhouse gases that allows parties to buy and sell permits for emissions (or credits) for reductions in emissions of specific pollutants

Emissions discharge of poisonous fumes

Fossil fuel fuel (such as coal, oil, and natural gas) that is formed inside the Earth over millions of years from remains of dead plants and animals

Glacier a slowly moving mass of ice formed by the accumulation and compaction of snow on mountains or near the Poles

Global warming potential ratio of heat trapped by a unit mass of a particular greenhouse gas to that of one unit mass of CO_2 during a specified time period

Global warming rise in temperature across the world that is believed to be caused by increase in the amount of certain gases (like CO_2) in the atmosphere

Greenhouse gas a gas that absorbs heat in the lower atmosphere of the Earth

Hurricane violent tropical storm fuelled by warm, moist air over ocean waters. Also known as cyclone or typhoon

Ice cap covering of ice over a large area

Landfill a large area, usually located in the outskirts of cities, where waste materials are buried under the ground

Methane a colourless, odourless greenhouse gas that can be used as fuel

Ozone layer a layer of ozone in the Earth's atmosphere that absorbs most of the ultraviolet radiation from the Sun reaching the Earth

Ozone a colourless, toxic gas with a pungent smell formed from oxygen that is a major air pollutant in the lower atmosphere

Radiation emission of energy as electromagnetic waves

Recycle to make something new from something that has been used before

Smog fog intensified by smoke or other atmospheric pollutants

Stratosphere upper layer of the Earth's atmosphere , which begins about 11 km above the Earth's surface and ends about 50 km above the Earth's surface

Sustainable involving methods that do not completely use up or destroy natural resources

Troposphere lowest part of the Earth's atmosphere extending from the Earth's surface to a height of about 6–10 km (lower boundary of the stratosphere)

Weather: the state of the air or atmosphere at a particular time and place

Test yourself!

1. Cultivation of rice leads to the emission of one of these gases.
- Methane
- CO_2
- Water
- Hydrogen

2. The molecular formula of Ozone is
- O_2
- O_3
- O_5
- $H_{2}0$

3. Which of these is also known as the laughing gas?
- Nitric acid
- Sulphuric acid
- Nitrous oxide
- Oxides of Nitrogen

4. What does the abbreviation HFCs stand for?
- Hydro fluoride caffeine
- Hydrogen fluoride capsule
- Hydrofluorocarbons
- Hydrogen fuelling carbons

5. **When is the International day for the preservation of Ozone layer celebrated every year?**
- 15 August
- 16 September
- 26 December
- 2 January

6. **Over which continent on Earth, a large, thin spot in the Ozone layer was first detected?**
- Australia
- Asia
- North America
- South America

7. **Which of these forests has been particularly affected by acid rain?**
- Black Forest, Germany
- Satpura Forest, India
- Amazon rainforest, South America
- Daintree Rainforest, Australia

8. **Which of these dates is associated with the beginning of Chipko movement?**
- 29 February 1974
- 26 March 1974
- 25 January 1978
- 28 February 1971

9. **The Montreal Protocol aims at resurrecting the depleted ozone layer between the years:**
- 2020–2030
- 2061–2065
- 2060–2070
- 2060–2075

Test yourself!

10. Where was ETS – Emissions Trading System – introduced?

- India
- China
- Pakistan
- USA

11. Who is the present Chairman of IPCC?

- Barack Obama
- Dr R K Pachauri
- Ban Ki-Moon
- Sonia Gandhi

12. What would be the Earth's average temperature, if the Greenhouse effect wasn't there?

- $18\,^0C$
- $-18\,^0C$
- $98.3\,^0C$
- $-98.3\,^0C$

13. The Ozone layer protects us from one of these rays.

- X-Ray
- Gamma Rays
- Beta Rays
- Ultraviolet Rays

14. Which of these GHGs or greenhouse gases was earlier used by dentists to give local anesthesia?

- Nitrogen dioxide or NO_2
- Nitrous oxide or N_2O
- Carbon dioxide or CO_2
- Carbon monoxide or CO

15. What replaced CFCs when their harmful impacts were discovered?

- GHGs
- HFCs
- NO_X
- VoCs